I0465162

cGMP Starter Guide

Copyright © 2016 www.validationresources.org

Author: Emmet Tobin

CONTENTS

1   What is GMP?   6

2   GxP, cGMP Explained
                                                7
3   Definitions of GMP
                                                8
4   Benefits of GMP
                                                11
5   A Brief History
                                                12
6   Consequences of GMP Failures   14

7   Quality Management System   19

8   Eight Principles of cGMP   23

9   What is Adulteration?   24

10 QMS- A Closer Look   25

11 The Aims of GMP – S-P-U-E   28

12 What Documents Require GMP?   30

13 Focus Areas of GMP   32

14 EU Regulations & GMP?   39

15 Audits   41

16 What is RFT?   44

17 What is PDCA?   45

18 What is 5S?   47

19   Counterfeit Protections   49

20   Good Documentation Practices   52

21   Why is GDP Essential?   53

22   Elements of GDP   54

23   Common Acronyms   65

24   Useful Definitions   66

ABOUT THIS BOOK

This concise book provides an introduction to Current Good Manufacturing Practices (aka cGMP). It introduces those who wish to work in regulated industries to GMP, highlighting key areas and practices. It is also a useful refresher for those with previous experience of cGMP.

1 WHAT IS GMP?

Good Manufacturing Practices are a set of practices that are required in order to comply with industry standards and regulations.

GMP helps to minimise the risks involved during manufacturing and helps to ensure products meet quality and regulatory standards.

A GMP quality system ensures that products are consistently produced and controlled according to predefined quality standards. It is designed to minimise the risks involved in any pharmaceutical production that cannot be eliminated through testing the final product.

2 GXP, CGXP
AND
CGMP EXPLAINER

Often, a broader term is used in industry -GxP-where the "x" is used as an umbrella letter representing different subjects or disciplines in industry. Some prime examples include GLP (Good Laboratory Practice), GDP (Good Documentation Practice), GEP (Good Engineering Practice) and GMP (Good Manufacturing Practices). Furthermore, the use of a lower case "c" as a prefix indicates "current" or "up-to-date". So cGMP stands for "Current Good Manufacturing Practices.

This means that some conventions or practices are subject to change within the industry. Therefore, it is important to be up-to-date in the application of cGxP or cGMP.

3 DEFINITIONS OF GMP

There are multiple regulators and organisations that provide definitions of "Good Manufacturing Practices". They include Organisations such as the World Health Organisation (WHO) and the International Society of Pharmaceutical Engineering (ISPE). Other definitions are offered by bodies such as the American competent authority for Food and Drug Administration. It is good to have an awareness of how organisations, bodies and competent authorities define GMP, and one should always review the "local" regulatory landscape. Below some definitions are provided to provide a feel for GMP and highlight the common thread between definitions.

W.H.O. World Health Organisation-"Good Manufacturing Practices (GMP, also referred to as 'cGMP' or 'current Good Manufacturing Practice') is the aspect of quality assurance that ensures that medicinal products are consistently produced and controlled to the quality standards appropriate to their intended use and as required by the product specification."

Food and Drug Administration: cGMP refers to the Current Good Manufacturing Practice regulations enforced by the US Food and Drug Administration (FDA). cGMPs ensure systems are properly designed and monitored, safeguarding the control of manufacturing processes and facilities. Adherence to the cGMP regulations ensures the identity, strength, quality, and purity of drug products by requiring that manufacturers of medications adequately control manufacturing operations. This includes establishing strong Quality Management Systems, obtaining appropriate quality raw materials, establishing robust operating procedures, detecting and investigating product quality deviations and maintaining reliable testing laboratories. This formal system of controls at a pharmaceutical company, if adequately put into practice, helps to prevent instances of contamination, mix-ups, deviations, failures and errors. This assures that drug products meet their quality standards.

MHRA (Medicines and Healthcare Products Regulatory

Agency) defines GMP as follows:

Good Manufacturing Practice (GMP) is that part of quality assurance which ensures that medicinal products are consistently produced and controlled to the quality standards appropriate to their intended use and as required by the marketing authorisation (MA) or product specification. GMP is concerned with both production and quality control.

4 BENEFITS OF GMP

Many of the drivers of GMP in effect are also benefits to the manufacturer. Good manufacturing practices are an expected practice in regulated industries and a manufacturer must meet all relevant GMP regulations if they wish to manufacture within a country or sell to a particular market.

It is important to maintain accurate, complete, up-to-date and consistent information to ensure patient safety and reduce any potential risks. Some benefits of GMP compliance and culture include:

GMP helps to reduce observations raised on inadequate documentation practices

GMP promotes efficiency and reduces the cost of doing business

GMP ensures products are safe for use in humans

GMP helps Pprevent/control contamination and cross contamination.

GMP reduces the risk of mislabelling and adulteration.

5 A BRIEF HISTORY OF GMP

1906: The Pure Food and Drug Act: Establishment of regulatory agency- FDA. The act makes it illegal to sell "adulterated" or "misbranded" food and drugs.

1938: Federal Food, Drug and Cosmetic (FD&C) Act: Sulfanilamide containing poisonous solvent causes 107 deaths.

1941: Insulin Amendment: requires FDA to test and certify purity and potency of insulin.

1962: Kefauver-Harris Drug Amendments Tragedy: Thalidomide tragedy causes severe birth defects in thousands of European babies.

1978: CGMPs Final Rule for drugs and devices (21 CFR Parts 210–211 and 820) establishes minimum Current Good Manufacturing Practices for manufacturing, processing, packing, or holding drug products and medical devices.

1979: GLPs Final Rule (21 CFR Part 58) -Good Laboratory Practices (GLP) for conducting non-clinical laboratory studies that support applications for research or marketing permits for human and animal drugs, medical devices for human use and biological products.

1982: Tamper-resistant packaging regulations issued for over-the-counter medicinal products: Acetaminophen-capsule poisoning by cyanide causes several deaths.

1987: Guideline on General Principles of Process Validation are issued by the FDA.

1996: Proposed Revision to US CGMPs for Drugs and Biologics (21 CFR Parts 21–211) adds detail for validation, blend uniformity, prevention of cross-contamination and out-of-specification results.

1997: Electronic Records Final Rule (21 CFR Part 11) requires controls that ensure security and integrity of all electronic data. 1998 Draft Guidance on Manufacturing, Processing, or Holding Active Pharmaceutical Ingredients and Investigating Out-of-Specifications and Test Results.

6 CONSEQUENCES OF GMP FAILURES

FDA Warning Lletters

This is an official message from the United States Food and Drug Administration (FDA) that usually states that it has found that a manufacturer or other organisation has violated some rule of the Quality System regulations.

FDA 483s

An FDA 483 letter typically includes a summary of findings and observations in relation to an audit or inspection where the FDA representatives have reason to believe GMP or other regulations have been violated or are not being met. In response to an FDA 483 letter, the company should address each item and provide a timeline for correction or request clarification of what changes are required.

Below are the top three items of concern:

- Procedures not written or are not fully followed
- Poor investigations of discrepancies or failures (CAPA)
- Absence of written procedures

Consent Decree

A consent decree is a binding order issued by a judge that stipulates the voluntary agreement by the participants in a case of litigation. Decrees are sometimes issued after one party voluntarily agrees to cease a particular action without admitting to any illegality of the action to date.

Product Recall

Product recalls can be initiated in circumstances where a manufacturer becomes aware of a manufacturing defect, packaging or mislabelling defect or otherwise. Product recalls can have a serious impact on the financial stability and future of a company due to bad publicity and loss of sales.

Plant Injunction

An injunction is a judicial process initiated to stop or prevent violation of the law, such as to halt the flow of violative products in interstate commerce and to correct the conditions that caused the violation to occur. (FDA 21 U.S.C. 332; Rule 65, Rules of Civil Procedure).

If a firm has a history of violations and has promised correction in the past but has not made the corrections, the injunction is more likely to succeed. However, the freshness of the evidence is critical.

For an injunction action to be credible in the eyes of the Department of Justice (DOJ), the U.S. Attorney and the court, the evidence must be current. Timeliness is an important factor when considering an injunction action, with or without a Motion for Preliminary Injunction or a temporary restraining order (TRO). However, case quality and credibility must not be sacrificed to meet guideline time frames. The purpose of the guideline time frames is to limit, as much as can reasonably be expected, the need to update evidence. Updating entails extra work at all levels of the case development and review process and more importantly, delays obtaining an injunction which is intended to stop violations that adversely affect the safety or quality of products in commerce.

When an injunction is granted, FDA has a continuing duty to monitor the injunction and to advise the court if the defendants fail to obey the terms of the decree. (FDA 6-2-Injuctions)

Debarment

The FDA has the authority to "disqualify," or remove, researchers from conducting clinical testing of new drugs and devices when the agency determines that the researcher has repeatedly or deliberately not followed the rules intended to protect study subjects and ensure data integrity. Further, the FDA can disqualify a clinical investigator who has repeatedly or deliberately submitted false information to the agency or study sponsor in a required report.

Under its statutory debarment authority, the agency may also ban, or "debar" from the drug industry individuals and companies convicted of certain felonies or misdemeanours related to drug products. Once individuals have been subjected to "debarment," they may no longer work for anyone with an approved or pending drug product application at FDA. Debarred companies may no longer submit abbreviated drug applications.

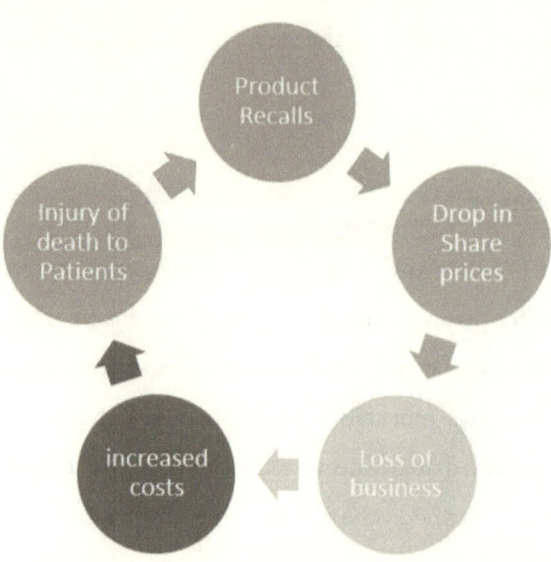

7 WHAT IS A QUALITY MANAGEMENT SYSTEM?

A Quality Management System, often abbreviated to (QMS) is any system based on a collection of business processes that are primarily focused on providing safe and quality products that consistently meet customer requirements. The core themes of a QMS are outlined below.

Customer and Regulatory Focus

An understanding of the customer needs and requirements should be evident within the organisation and with the future vision of the company. The company should have an understanding of the regulatory landscape as this is subject to change over time. In turn the company should be positioned to respond to that change.

Leadership

To truly lead, one must be accepted in the hearts and minds of those they lead. Authentic leadership pays off. A leader should foster a sense of togetherness and common vision. A leader is anyone who influences or directs people either formally or informally. We are all leaders to some extent.

Involvement

Engagement by everyone across an organisation is now recognised as being key in the successful deployment of any Quality Management System. Everyone should have a voice within the company. As the saying goes "we are only as strong as the weakest link" is very apt.

The Process Approach

ISO 9001 and ISO 13485 are standards that are based on process approaches. A process approach essentially utilises methods or standardised tools to help drive consistency.

Some common tools include DFMEA /DMAIC/PDCA to name but a few.

Systems Management

This essentially means that systems are defined and described in writing along with the appropriate responses to expected issues that arise. Effective systems management must ensure that the various systems work in support of each other and communicate effectively with one another.

Decision Making

In order to make the right decision, the person empowered to make the decision must be informed. To be correctly informed one must have the correct details and facts available. In a manufacturing environment the facts are essentially data and the analysis of data.

During manufacturing or processing, data is generated as a result of monitoring and measurement of products and the related processes.

Supplier Management

Don't ruffleyour suppliers' feathers. Security of supply is key in delivering products to customers or patients again and again, Raw materials or sub-components sourced from external suppliers must always be sourced at the right price and time with the emphasis on getting the best quality possible.

Continuous Improvement

For ISO 13485 continuous improvement refers to improving the effectiveness of the Quality Management

System. It is harder to drive improvement of the product due to regulatory and practical requirements.

This is a key difference in contrast to ISO 19001:2008 as there is a requirement to continually improve both product and processes.

8 EIGHT PRINCIPLES OF CGMP

1. Design and construct the facilities and equipment properly with due regard for regulations and international ISO standards.

2. Write and follow written procedures and instructions at all times.

3. Document work accurately and in a timely fashion.

4. Validate equipment systems and processes.

5. Monitor facilities and equipment.

6. Protect against contamination and maintain a clean environment.

7. Control raw materials, components and product related processes.

8. Conduct periodic audits.

9 WHAT IS ADULTERATION?

If there is a failure to comply with cGMP regulations, a regulatory authority can level the charge of "adulteration" in respect of certain products.

A drug product is deemed to be adulterated if the methods used in or the facilities or controls used for its manufacture, processing, packing or holding do not conform to or are not operated or administered in conformity with cGMP to ensure that such drug meets the requirements regulations and has the identity and strength and meets the quality and purity characteristics which it purports or is represented to possess. (Source: www.fda.gov).

10 QUALITY MANAGEMENT SYSTEMS- A CLOSER LOOK

The key elements of a QMS are listed below. The ISO Standard, ISO 9001, is a global Quality Management standard used by thousands of organisations and companies. This standard sets out the requirements of a QMS.

Quality Policy

A company will document their commitment and approach to quality within their organisation. It usually sets out how they plan to achieve a high and consistent standard of quality. It should in some way speak to the customer or end user.

Quality Objectives

Quality objectives can be documented in a Quality Plan at site or organisational level. An effective way of defining quality objectives is use of the SMART method. SMART stands for Specific, Measureable, Achievable, Realistic and Timely.

Quality Manual

An in-house guidance document to provide a framework for achieving the quality objectives.

Organisational Structure and Responsibilities

Organisational charts can be used to map out the company structure. Roles and responsibilities can be documented in site quality plans, job descriptions and Standard Operating Procedures.

Data Management

A coherent approach to the provision, storage and maintenance of data.

Processes

Processes are defined and documented.

Resources

Resources must be properly understood, allocated and linked across the organisation.

Product Quality & Customer Satisfaction

The proper management and investigation of complaints is

important to reduce future instances from reoccurring. Continual engagement with the end user or customer is critical.

Continuous improvement including corrective and preventive action- where continuous improvement projects and initiatives are encouraged and supported. The application of a CAPA system to ensure quality is maintained and consistent.

Maintenance

A Preventative Maintenance schedule is in place and managed accordingly.

Sustainability

All work practices are sustainable and consistent throughout the lifecycle of processes and products.

Auditing

Systems are auditable and maintained to allow internal or external review and audit.

Engineering Change Control

Where changes are required to validated processes or equipment, changes are managed and introduced under change control.

11 THE AIMS OF GMP

S-P-U-E

A common acronym used to highlight the aims of Good Manufacturing Practices (GMP) is SPUE which stands for Safe-Pure-Uniform-Effective. This definition is particularly suited to pharmaceutical products as the chemicals and drugs used need to be pure and free of contaminants. Furthermore, they need to be uniform, meaning they will have the same constituents from tablet to tabletand batch to batch. A description of each word is shown below:

SAFE- the product has the right ingredients if it is a drug product. It is packaged as intended and correctly labelled in order to provide identification and safe use.

PURE- it is free of contaminants, foreign matter, chemicals and harmful microbes.

UNIFORM- The product is manufactured consistently and will have the same quality between batches manufactured on different days.

EFFECTIVE- Ultimately, the product must be effective in treating the medical condition. To be effective, it requires the correct ingredients, the correct amount of ingredients and correct packaging to maintain the product stability over time.

12 WHAT DOCUMENTS ARE SUBJECT TO GMP?

Within a regulated company, many documents are used to instruct, track, test and record information on the manufacturing process. Any document that can impact the quality of the product or product safety is treated as a controlled document. A controlled document is classified as a legal document. These controlled documents must incorporate certain requirements such as the date of approval, revision control and appropriate levels of review and approval.

The accuracy and content of these documents can be subject to review by regulatory bodies Including the FDA in the US and theMHRA in the UK

It is important that there are no errors or "questions marks" over the content. Examples of controlled documents include:

- Policies
- SOPs
- Specifications
- MFR (Master Formula Record)
- BMR
- Validation protocols and reports
- Forms
- Logbooks
- Records
- Bills of Materials (BOMs)
- Test Methods

13 FOCUS AREAS OF GMP

#01 Written Procedures

Written procedures are controlled documents that provide detailed step-by-step instructions for the user. Written procedures promote consistency as they allow the same task to be performed in the same way, even by different people.

They also act as a reference. If changes or improvements are identified, having a procedure in place creates a clear starting point which can be improved or modified in a controlled manner.

-Procedures should be written using clear and concise language

-Steps should be numbered clearly and individually to make them easy to follow

-Remember, written procedures are only effective if they are followed correctly, consistently and at all times by everyone

Never deviate from written procedures, these controlled documents ensure consistency and accuracy is maintained over time

#02 Training

When personnel are trained in written procedures and they are applied consistently, this reduces the risk of mix-ups and errors in the manufacture, testing and packaging of products.

Adhering to written procedures is a requirement of GMP regulation. Even though an individual may believe they have a faster or different way of completing a job, the approved written procedure must be followed to maintain consistency.

If you or an employee has a suggestion to improve an SOP or process, then raise the issue with your manager or quality representative.

Hierarchy of Documentation

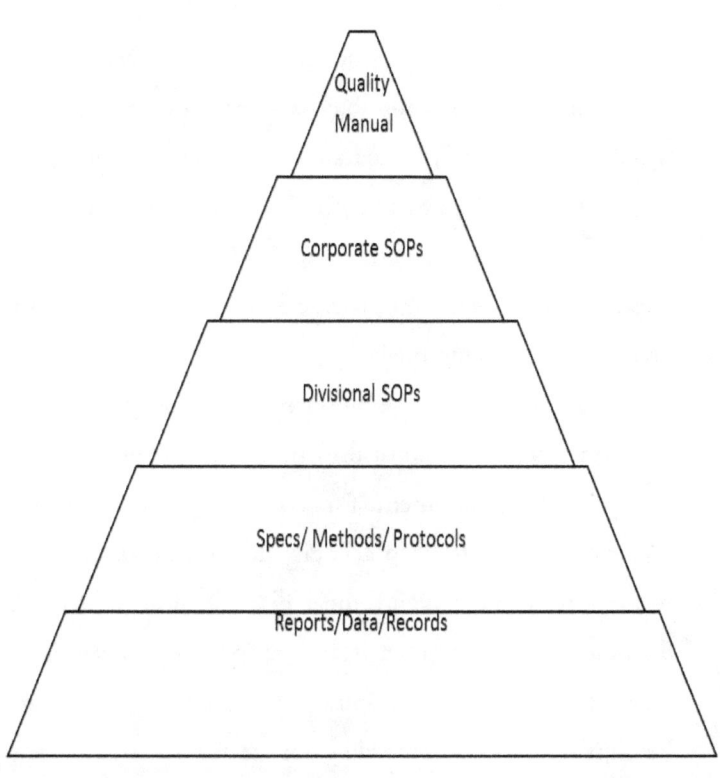

#03 Documenting Work

As the saying goes "Good Science starts with good documentation", meaning that data should be recorded as soon as available and accurately, following principles of GDP (Good Documentation Practices). The rule of thumb "If it's not documented, it didn't happen" is an important statement. To provide evidence or proof that a process is producing quality product, it must be verifiable in a document. Documentation requires that you record, sign and date every step of the operation you perform. It is a regulatory requirement for companies manufacturing medical products to keep accurate documents and records relating to products and their manufacture. To ensure documents meet a high standard of compliance both prompt recording and accurate recording is paramount. Prompt recording- recording of test results and data should take place as soon as possible and as soon as available. This is to minimise risk or errors with late entries which could lead to quality issues or audit findings.

#04 Systems and Processes

Validation is defined as "The collection and evaluation of data, from the process design stage through commercial production, which establishes scientific evidence that a process is capable of consistently delivering quality products." In essence, validation provides proof that you can manufacture products to specifications again and again, day after day, from operator to operator.

It is a GMP requirement to demonstrate and document control of the critical aspects of certain operations. Validation must be completed for equipment, systems and processes used in the manufacture of medical devices and regulated products.

#05 Cleanliness

The purpose of maintaining a clean manufacturing environment is to prevent or minimise the contamination of products. The level of cleanliness is dependent on the type of products being manufactured. For example, a sterile intravenous product would require a higher degree of cleanliness and higher environmental controls. Each manufacturing plant or company should have a Plant Sanitation Program a.k.a. Environmental Monitoring Programme.

Principles of Good Hygiene:

• Change clothes on a daily basis

• Wear appropriate clothing

• Wash hands thoroughly, according to company procedures

• Report any spillages

• Inform your manager of illness

• Check work area when commencing and concluding work

• Carefully follow all cleaning schedules and written sanitation procedures

• Prevent the entry of rodents, insects and other pests by keeping doors closed

• Ensure waste is disposed of safely

• A culture of responsibility and attention to cleanliness

• Routinely clean manufacturing equipment, replace filters and clean drains and drain trays

#06 Quality

The key to quality is consistency. If you manufacture a quality product consistently, then your customers will return and have a positive experience of your company and its products. Quality is providing a high degree of assurance that products will be defect free, effective and safe. Effective controls work to ensure that a company delivers quality time and time again. These include;

- Control of raw materials and components

-Control of the manufacturing process

-Packaging and labelling controls

-Release of product

-Parts received from supplier are inspected

-Most suppliers provide Certificates of Conformance (CofC) stating that the materials meet specifications -- Components are labelled appropriately and securely stored

-Rejected components are quarantined pending investigation or disposed of safely

14 WHAT EU REGULATIONS RELATE TO GMP?

The European Directive "2003/94/EC for medicines and investigational medicines for human use" is the core legislation that applies in the EU. Compliance to this directive is mandatory for manufacturers and it enables companies to legally manufacture and sell their products within the European Economic Area.

EudraLex - Volume 4 Good manufacturing practice (GMP) Guidelines: Volume 4 of "The rules governing medicinal products in the European Union" contains guidance for the interpretation of the principles and guidelines of good manufacturing practices for medicinal products for human (and veterinary) use.

Differences between EU & US Regulations

- In the US, it is mandatory to use the code of federal regulations, while in Europe the key document for GMP inspections is the European Commission Directive 2003/94/EC.

- The EU directive contains a series of supplementary annexes. The US does not.

15 AUDITS

Simply put, an audit is a review activity that examines if a company or organisation's processes are being followed. It also allows the identification and improvement of any concerns or non-compliances. Audits can be internal (conducted by internal staff) or external, (external-regulatory audits or certification bodies).

Audits are a key element of a Quality Management System. The process of establishing an internal audit process can be aided with reference to ISO 19001. This standard provides guidance and lots of examples on implementing and maintaining audit systems.

Audits provide a means of assessing a company's Quality Management System and how well it is in compliance with the processes and procedures within the company.

Some Key Benefits of Audits:

Audits help verify compliance and conformity to requirements laid out in regulations and industry legislation e.g. ISO, FDA, EudraLex etc.

They measure the effectiveness of the QMS and the engagement of top management.

Audits help identify opportunities for improvements.

Audits promote awareness of the Quality Management System.

Pre Audit Questions

Prior to any audit, the scope and criteria should be clearly defined and documented. Some key points are examined below.

Scope: the scope of the audit should be clear. This should define the products, processes and location on which the audit will be based.

Audit Criteria: the criteria should be based on the requirements of the standard to which conformity is sought after. Typical standards include ISO 9001 (General Quality Management Standard), ISO 13485 (Medical Devices) and FDA 21 CFR Part 820 (Quality System Regulation).

Audit Objective: An audit may be necessary in order to maintain certification. It may also just act as an opportunity to identify potential improvements of the QMS.

16 WHAT IS RFT?

Right First Time strives to create a culture of excellence. People are challenged with performing their tasks always in the correct manner to achieve the correct results always - *right the first time.*

RFT is the enabler to providing customers worldwide with accessible, high quality and advanced healthcare solutions which comply with cGMP requirements.

RFT in Practice
Achieve excellence *rather than* "that's good enough"
Prevent defects *rather than* "detect defects"
Right first time *rather than* rework

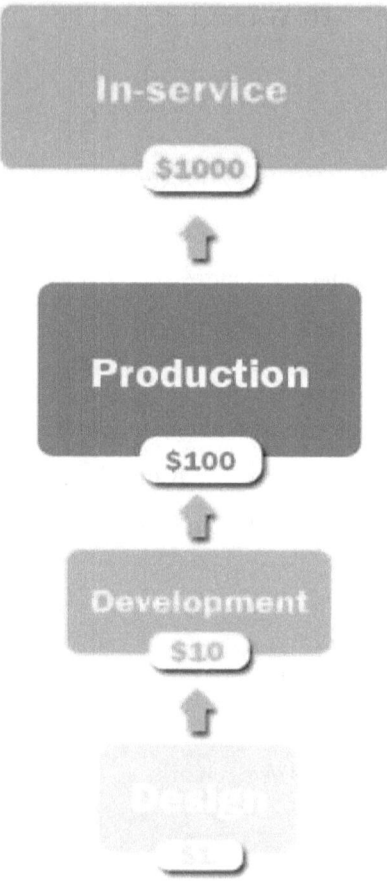

17 WHAT IS PDCA?

PDCA (Plan–Do–Check–Act) is a four-step management tool often used in GLP and GMP environments

It sets in motion a repeatable and structured process-driven approach to solving problems and helps to drive consistent practices.

Plan

The Plan step is used to establish the objectives and desired goals of the proposed changes or modifications. Documenting these goals is important as it will drive the aspects of the next steps in the PDCA Process.

Do

Implement the plan and the changes identified. The "Do" step may require data collection and/or analysis prior to the implementation of changes. Training may also be required. Responsibilities should be clearly defined.

Check

Review results and analysis against the planned and expected results or goals.

Act

The act step ensures if any further corrective actions or modifications are noticed in the check step. The processes will require the person to "act" on the findings. However, any proposed changes are better captured by returning to the first step and restarting the process, either way, the application of PDCA will drive continuous improvement.

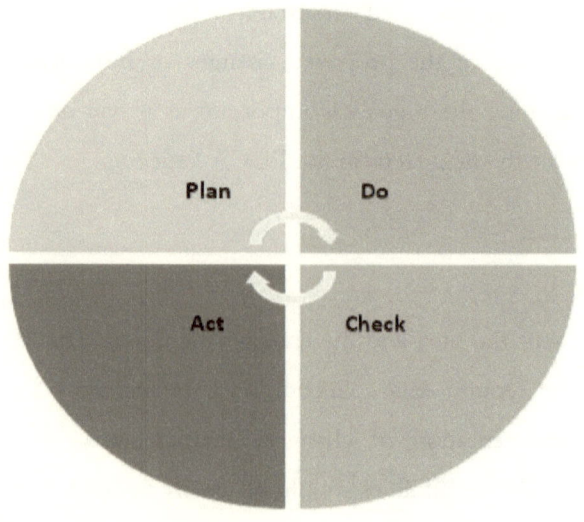

18 WHAT IS 5S?

5S is a Japanese methodology of organising and storing items in a work or lab environment. It has been adopted by many Western companies as a tool to help maintain standards and reduce errors and mix-ups. The "5s" represents each stage of the method.

Sort

Sorting out any items that are not in use and removing to a more appropriate area or to storage or the bin.

Set-in-Order

The idea of "Set-in-Order" is to be alwas organised. "A place for everything and everything in its place. "If we "set-in-order" we can help to make live processing and testing more efficient and reduce the risk of errors, omissions and accidents.

Shine

Regular cleaning is an important practice and it is always

helpful to "Clean as you go."

Standardise

Implement standard practices through SOPs and training. Standardisation can also be applied to work station layout.

Sustain

Make it a habit! After implementing a 5s methodology, it is only effective if continuous efforts are made to "sustain" the changes.

Sort- Set-in-Order- Shine – Standardise – Sustain

19 COUNTERFEIT PROTECTIONS

The definition according to the WHO of a medical counterfeit product is *"when there is a false representation in relation to its identity (e.g. any misleading statement with respect to name, composition, strength, or other elements), its history or source (e.g. any misleading statement with respect to manufacturer, country of manufacturing, country of origin or the marketing authorisation holder)."*

This definition applies to the product, the container and other packaging or labelling information. Counterfeit products may include:

products with the correct active ingredient(s), in the correct proportions;

products with the correct active ingredient(s), but with the wrong dosage;

products without active ingredient(s);

products with impurities or toxic ingredients.

Packaging Technology Designed to Reduce Counterfeiting

DataMatrix: In January 2011, traceability technology based on bar codes (DataMatrix) was put in place in Europe. The ultimate objective and purpose of the DataMatrix is to help ensure traceability of each box in its supply chain to the pharmacist until the end-user, the patient.

Unique Device Identification (UDI) System for Medical Devices: The purpose of the (UDI) is intended to assign a unique identifier to medical devices sold within the United States. It was signed into law on September 27, 2007, as part of the Food and Drug Administration Amendments Act of 2007. The unique identifier enables identification of the device through distribution and use.

A secondary benefit of the UDI is it helps to identify counterfeit products by improving the ability to distinguish between authentic and counterfeit devices.

20 GDP

This chapter provides a comprehensive and easy-to-understand guide to the subject of Good Documentation Practices. Good Documentation Practices (commonly abbreviated to GDP or GDocP) is a term used to describe standards by which documents are created, modified and maintained.

GDP is a practical skill that is required within the life science sector (medical device, pharmaceutical and so on). It can be broadly divided into two streams; GDocP practices and how they apply to electronic document and secondly, GDocP for handwritten entries including initial and dating and recording of data and test results by hand.

GDocP is fundamental in achieving compliance to Good Manufacturing Practices (GMP). It is required in the U.S. by the FDA's Code of Federal Regulations and in Europe by the governing body EudraLex.

If GDocP is not practiced it jeopardises the integrity of data and written records. This can lead to to the falsification of data which is a serious regulatory offence.

21 WHY GDP IS ESSENTIAL?

Admittedly, implementing and maintaining GDP takes time, effort and resources, however, there are some benefits that come with it. Most importantly, Good Documentation Practice is an expected practice in regulated industry as trust and ethics are fundamental to business.

It is important to maintain accurate, complete, up-to-date and consistent information to ensure patient safety and reduce any potential risk to patients.

Practising GDP equally helps to reduce observations raised on inadequate documentation practices at times of audit by regulated bodies such as the FDA.

It helps to improve communication and efficiency within companies. If GDP is not followed it can call into question other processes and procedures within a company.

22 ELEMENTS OF GDP ?

Documentation Creation

The principles of GDP should be applied at the document creation stage. As most people are familiar with softcopy or electronic documents, some of these points are obvious but nonetheless need to be made. All documents should be electronically written and not handwritten except for execution of protocols, test results and adding entries. Documents that are approved controlled should be:

Accurate and free from errors

Have revision or version controlled

Should have an effective date or date of release

Approval of Documents

Document approval must be completed by trained and appropriately experienced personnel. Often companies will use an approval matrix which explains which people are required to approve each document. For example, an EHS (Environment Health and Safety) officer would be required to approve a risk assessment.

Signatures

A signature on any document is legally binding so remember to read and understand what is being signed for. Every signature should also include the date in the correct format. If a signature appears within the same document alongside initials, substituting a full signature with initials and date is generally acceptable. This practice is common when large documents are being completed.

Date and Time Format

A standardised approach to dates and times is important especially within large global organisations. For instance, in the USA, the norm is to place the month before the date, whereas in Ireland and Great Britain it is common to write the day of the month followed by the month. Most companies would define their date and time format in an SOP or procedure.

The date and time format can also be configured in Word documents and Excel worksheets to align with a companies preferred date and time format.

Handwritten Entries

When a handwritten entry is required such as a signature or a test result, indelible ink must be used. Many companies will have an SOP or procedure that states the specific ink colour required.

Remember - Never use pencil when making a handwritten entry, always use indelible ink

If an entry of a test result or test data isn't completed at the time of execution, this constitutes a late entry. Backdating an entry or signature is forbidden. Always use the correct and current date.

How Are Mistakes Corrected?

This is a critical area of GDP. Failure to follow the requirements of GDP when correcting mistakes is the most common failure when it comes to documentation in industry. The method of correcting mistakes using GDP allows for a person looking at the document for the first time to clearly see the original entry and the corrected entry. This maintains the integrity of the document. In order to identify the changes and corrections, certain rules must be followed. No overwriting is allowed and white-out or Tipp-Ex is not allowed. The correct way to make any changes or corrections by hand is shown in the diagram below.

Accuracy

Accuracy of information provided in documents is critical in the life science industry. As the end user is a patient, inaccurate records or documents could cause serious injury or death. Controlled documents are also legal documents and could be called upon if recalls, litigation or investigations arise.

Many documents used in the manufacture of medical devices are designed to record information or test results. These test results are then used to disposition (pass or fail) batches of product. Inaccurate information could risk the release and distribution of defective product. This has a potential impact on both the business and the patient or user.

Blank Spaces or Blank Fields

On completion of a document such as a logbook or record, no blanks spaces should be left unfilled. This is to avoid late entries and also to prevent confusion. Blank spaces or blank fields should have a diagonal line drawn neatly across the space, the letters "N/A" written and the entry signed and dated. If the reason for "N/A" is not evident then it is wise to include an explanatory note or sentence.

Data Transcription

Transcribing is the process of transferring data from one source to another. This is often required when raw data is involved. When data is in raw format it may need to be entered into a Microsoft Excel sheet. When transcribing data remember that all original raw data must be stored in case it is needed at a future date.

After the data is transcribed it must be verified by a second person to check for any errors or omissions.

The individual responsible for data verification should:

Be independent of the data

Have the technical knowledge to spot errors

Communicate to the author/recorder if errors are present

Ensure errors are corrected and meet GDP requirements

Revision Control

Controlled documents should always have a version number or revision number electronically on each page of the document. This is similar to books which always list what edition they are e.g. first edition or second edition.

Revision control is a key element of the Quality Management Systems in place in regulated industries. As the need for changes in the document arises, the controlled document can be amended/updated. With each update the version number revises also. Some companies will use alphabetic revision control and to a lesser extent numeric revision control (Version A, Version B or Version 01, Version 02).

Management of Attachments

Attachments to controlled documents can include training records, data sheets, lab results and so on. It is important that attachments are identified for traceability purposes. If the attached becomes detached from the main document, the attachment should be identifiable.

It is best practice to include a reference number on the attachment if available. If the attachment consists of several pages, each page should be numbered in Page X of Y format if not electronically done so. And remember, hand written entries must be accompanied by a signature and date.

Always use staples to attach documents together. Glue or paper clips are not acceptable.

Management of Documents through Their Lifecycle

GDP applies to all the different stages of a document's lifecycle. These stages include creation, review, approval, issuing, completion of records, revision, updating, retirement and storage. Storage a.k.a. retention is an important stage and often a legal requirement for medical devices and pharmaceutical products. For consumer OTC medicines a 5-year retention of quality records often suffices. For implants such as TKRs or Total Knee Replacements, a 90-year retention period is required. This ensures that traceability and a quality record is available if the need arises.

Test Results

This section provides an overview on the correct handling of test results. Test results can be generated from various types of product testing such as visual inspection, dimensional inspection and chemical analysis. The recording of all test results should be completed on an approved form. This is to ensure that the correct information is being recorded and the same approach is taken by different people who might have to complete testing.

Calculations

There are different ways calculations can be completed. Many simple calculations can be done by an individual using a calculator, alternatively, a software package such as Minitab or an Excel sheet can be used to complete complex calculations. It should be clear to the reader what calculation is required, what the formula is and how the calculation is completed.

If the formula used is not included on the sheet, it should be referenced in a controlled document.

Care is also required when recording numbers of several decimal places in length, as rounding error can be introduced.

Units of Measurement

The most important thing to remember is consistency in units of measurement when recording data or making calculations. Consult your company procedure if available to determine the correct units of measurement.

Many U.S. companies use imperial units e.g. inches, pounds etc. In Europe the International System of Units or SI is used, e.g. millimeters and kilograms.

23 COMMON ACRONYMS

CFR: Code of Federal Regulation

EMEA: European Medicines Agency

FDA: Food and Drug Administration (US regulatory body)

HPRA: Health Products Regulatory Authority

cGMP: Current Good Manufacturing Practice

24 USEFUL DEFINITIONS

Validation: confirmation via documented evidence that the particular requirements for a specific intended use can be consistently fulfilled under anticipated conditions.

Verification: confirmation by examination and provision of objective evidence (i.e. documentation) that the specified requirements have been fulfilled.

cGMP: Current Good Manufacturing Practices.

Concurrent Validation: Validation activities occurring at the same time as one another or concurrent to a product launch.

Prospective Validation: This is when validation is done *in advance of commercial manufacturing.*

Retrospective Validation: Retrospective validation is used for facilities or processes that have not completed formal validation. Historical data or a retrospective review can provide the evidence that the process or facility is *operated as intended.*

Standards: (ASTM, ISO etc.) is a document standard that provides requirements, specifications and guidelines to achieve consistency. The term consistent can refer to materials, products and processes to ensure they are fit for the intended purpose or use. Standards are not always mandatory. However, they help manufacturers be compliant with directives/legislation. Standards also represent the current and best practice in the field of study/industry.

Directives: Directives are legal requirements. These must be met by manufacturers. Standard such as ISO 13485 help companies meet the requirements of directives, such as "Guidelines Relating to the Application of the Council Directive 93/42/EEC on Medical Devices."

Notified Bodies: A notified body is a certification organisation which the national authority (the competent authority) of a member state designates to carry out one or more of the conformity assessment procedures or audits described in the annexes of the medical devices directives or GMP legislation.

Competent Authority: When it comes to medical devices,

a competent authority is the legally designated authority mandated to monitor compliance with directives and legal requirements within the industry. The competent authority has the power to grant and revoke licences.

REFERENCES

GAMP 5 – A Risk-Based Approach to Compliant GxP Computerized Systems

FDA 21 CFR Part 11

FDA 21 CFR Part 820

Medical Device Directive (93/42/ECC)

EN ISO 13485:2003

ANSI/AAMI/ISO 13485:2003 - Medical devices – Quality management systems, Requirements for regulatory purposes